P·I·C·T·U·R·E·P·E·D·I·A

NOTE TO PARENTS

This book is part of PICTUREPEDIA, a completely
new kind of information series for children.
Its unique combination of pictures and words
encourages children to use their eyes to discover and
explore the world, while introducing them to a wealth
of basic knowledge. Clear, straightforward text
explains each picture thoroughly and provides
additional information about the topic.

"Looking it up" becomes an easy task with
PICTUREPEDIA, an ideal first reference for all types of
schoolwork. Because PICTUREPEDIA is also entertaining,
children will enjoy reading its words and looking
at its pictures over and over again. You can encourage
and stimulate further inquiry by helping your child
pose simple questions for the whole family to
"look up" and answer together.

SPACE

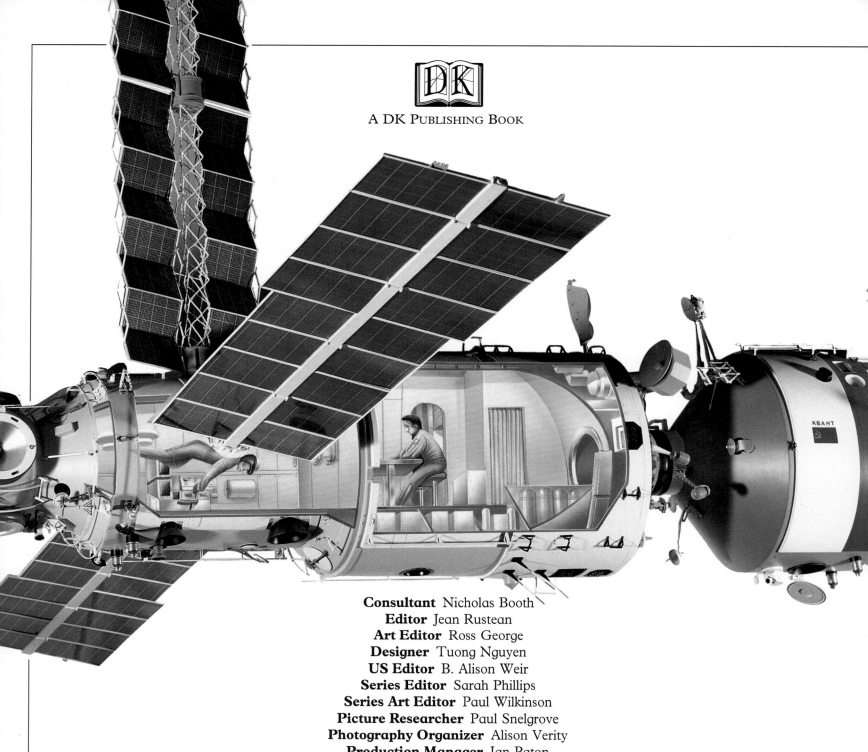

DK

A DK PUBLISHING BOOK

Consultant Nicholas Booth
Editor Jean Rustean
Art Editor Ross George
Designer Tuong Nguyen
US Editor B. Alison Weir
Series Editor Sarah Phillips
Series Art Editor Paul Wilkinson
Picture Researcher Paul Snelgrove
Photography Organizer Alison Verity
Production Manager Ian Paton
Editorial Director Jonathan Reed
Design Director Ed Day

First American Edition, 1992
10 9 8 7 6 5
Published in the United States by
DK Publishing, Inc., 95 Madison Avenue
New York, New York 10016

Library of Congress Cataloging-in-Publication Data
Space / Nicholas Booth, editor. – 1st American ed.
p. cm. – (Picturepedia)
Includes index.
Summary: Examines different aspects of outer space and our efforts
to study and explore it, including constellations, voyages to the
moon, the planets of the solar system, and galaxies.
ISBN 1-56458-141-1
1. Astronomy – Juvenile literature. 2. Astronautics in astronomy –
Juvenile literature. 3. Outer space – Juvenile literature.
[1. Astronomy. 2. Outer space.] I. Booth, Nicholas. II. Series.
QB46 S689 1992
520–dc20
                                                    92–52837
                                                    CIP
                                                    AC
Reproduced by Colourscan, Singapore
Printed and bound in Italy by Graphicom

SPACE

DK PUBLISHING, INC.

CONTENTS

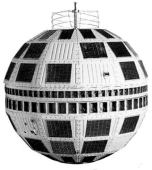

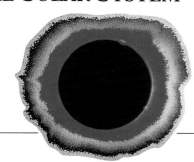

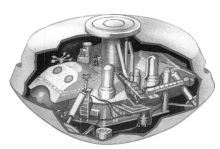

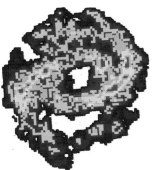

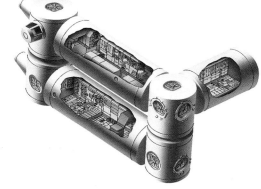

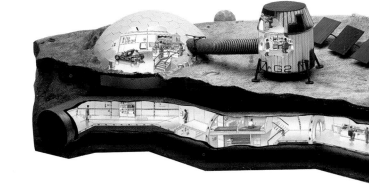

INTO THE UNIVERSE

People have always been curious about the things they could see up in the sky. On a clear night, it is possible to see the Moon and hundreds, even thousands, of stars. People who study the stars and planets are called astronomers. The universe is made up of galaxies, stars, planets, moons, and other bodies scattered throughout space.

Comet

Supernova

Sky Watcher
This is Galileo Galilei. He lived in Italy about 350 years ago. He was one of the first people to use a telescope to study the Moon and planets. He proved that the Earth moves around the Sun.

Arrows shot from a fire basket

Up into Space
The first rockets were invented in China more than 800 years ago and worked by using gunpowder, like fireworks today. Years later, rockets were made that could travel fast enough to take people into space.

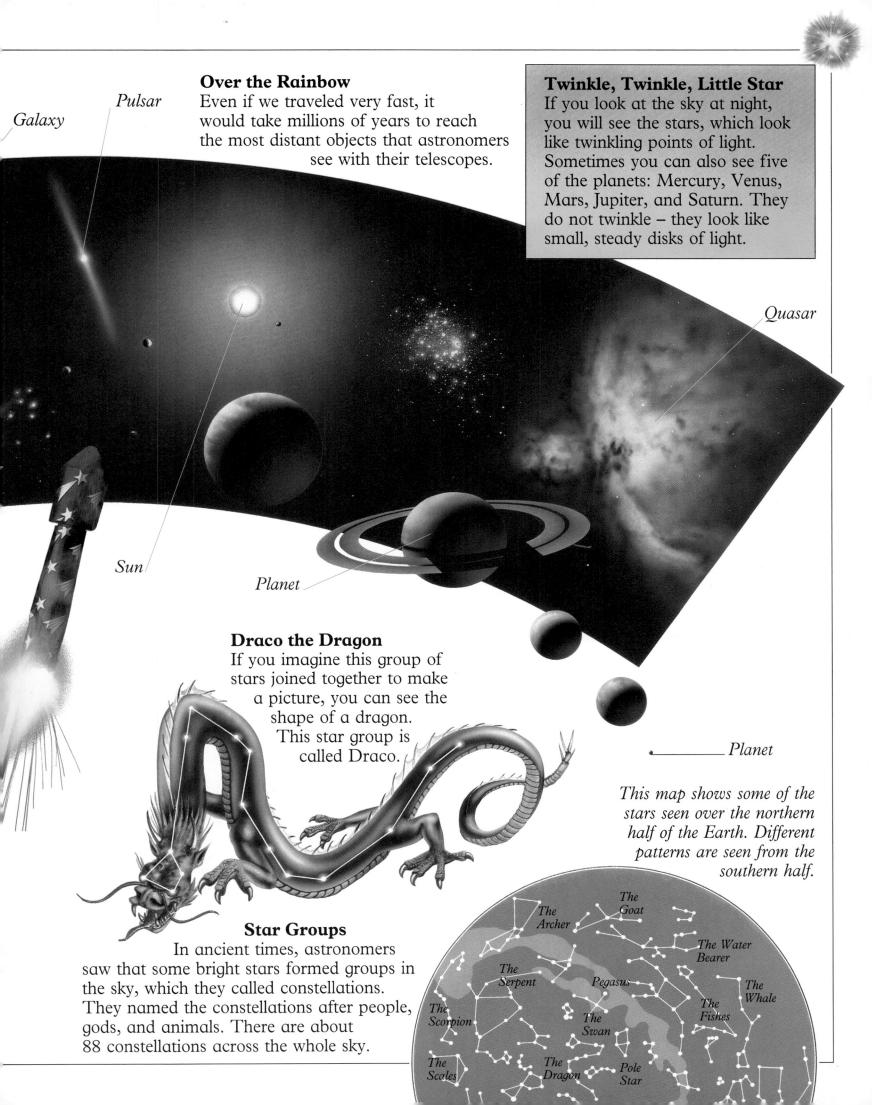

Galaxy

Pulsar

Over the Rainbow
Even if we traveled very fast, it would take millions of years to reach the most distant objects that astronomers see with their telescopes.

Quasar

Sun

Planet

Draco the Dragon
If you imagine this group of stars joined together to make a picture, you can see the shape of a dragon. This star group is called Draco.

Planet

This map shows some of the stars seen over the northern half of the Earth. Different patterns are seen from the southern half.

Star Groups
In ancient times, astronomers saw that some bright stars formed groups in the sky, which they called constellations. They named the constellations after people, gods, and animals. There are about 88 constellations across the whole sky.

The Goat

The Archer

The Water Bearer

The Serpent

Pegasus

The Whale

The Scorpion

The Swan

The Fishes

The Scales

The Dragon

Pole Star

CONSTELLATIONS

The groups of stars we see in the sky are called constellations. These groups have Latin names, such as Ursa Major, which means Great Bear. When you first go outside to look for a constellation, you may find it difficult to spot among all the stars in the sky. The Earth is spinning around very slowly – so that over a few hours it may seem as if the stars have moved across the sky, but if you gaze for long enough, you can pick out the patterns made by the brightest stars.

Leaping Lion
If you imagine the stars in a group have been joined to make a picture, you will see why this constellation is called Leo, the Lion.

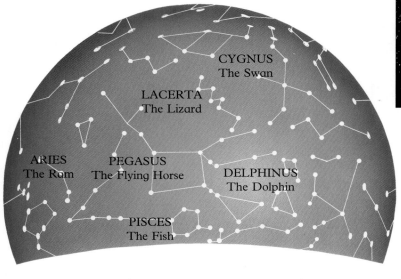

Northern Hemisphere

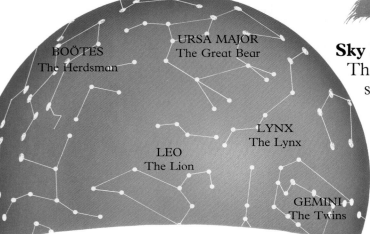

BOÖTES
The Herdsman

URSA MAJOR
The Great Bear

LYNX
The Lynx

LEO
The Lion

GEMINI
The Twins

Sky Lights
These two pictures show what the night sky would look like if you were standing, looking up, in the northern half of the world at one time of the year.

Light-years
Huge distances are often measured by astronomers in light-years. A light-year is the distance light can travel in a year. Light takes eight minutes to reach the Earth from the Sun. Light from the next-closest star takes more than four years!

CYGNUS
The Swan

LACERTA
The Lizard

ARIES
The Ram

PEGASUS
The Flying Horse

DELPHINUS
The Dolphin

PISCES
The Fish

Orion, the Hunter
The constellation Orion can be seen from most parts of the world. The three stars together form Orion's belt. Below the belt is his sword.

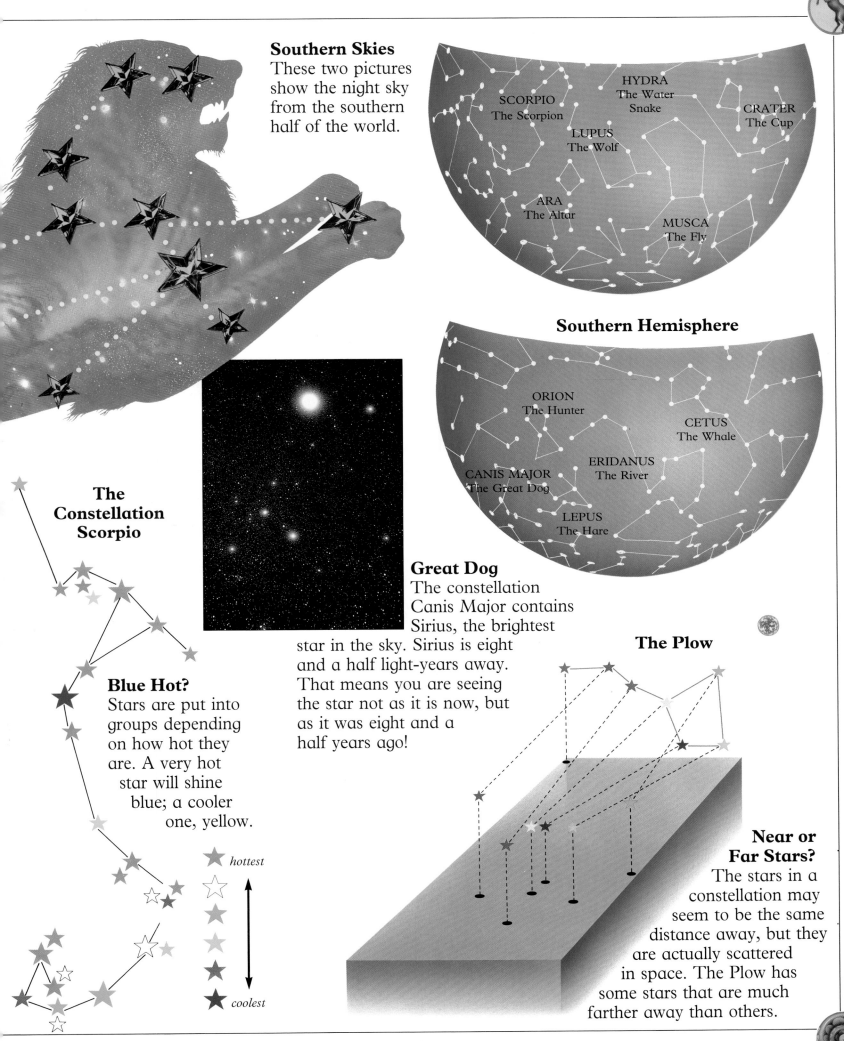

Southern Skies
These two pictures show the night sky from the southern half of the world.

SCORPIO
The Scorpion

HYDRA
The Water
Snake

CRATER
The Cup

LUPUS
The Wolf

ARA
The Altar

MUSCA
The Fly

Southern Hemisphere

ORION
The Hunter

CETUS
The Whale

ERIDANUS
The River

CANIS MAJOR
The Great Dog

LEPUS
The Hare

The Constellation Scorpio

Blue Hot?
Stars are put into groups depending on how hot they are. A very hot star will shine blue; a cooler one, yellow.

hottest

coolest

Great Dog
The constellation Canis Major contains Sirius, the brightest star in the sky. Sirius is eight and a half light-years away. That means you are seeing the star not as it is now, but as it was eight and a half years ago!

The Plow

Near or Far Stars?
The stars in a constellation may seem to be the same distance away, but they are actually scattered in space. The Plow has some stars that are much farther away than others.

LIFTOFF

Rockets were invented in China a long time ago. They looked a bit like arrows and worked by burning gunpowder that burns up very quickly, so the rockets did not travel very far. Since then, people have tried many ways of sending rockets up into space. In modern rockets, two liquid fuels are used. They mix together and burn. Then the hot gas shoots out of the tail, pushing the rocket up and away.

V-2 Rocket
1945

Gemini
Titan 1964

The Fly!
In 1931, a German named Johannes Winkler launched his HW-1 rocket. It went 7 feet (2 m) into the air, turned over, and fell back to the ground. A month later, he tried again. This time it climbed to 297 feet (90 m) and landed 660 feet (201 m) away.

3,2,1, Fire!
A hundred years ago, soldiers used rockets like this. They were called Congreve rockets.

Saturn Power
Saturn 5 is the biggest rocket ever built. It is as tall as a 30-story building! It carried the first American astronauts to the Moon.

Fuel tank

The stabilizing fins keep the rocket on course.

Five rocket engines

See It Go!
If you blow up a balloon and let it go without tying a knot in the neck, the air will rush out very quickly. When the air goes out one way, it pushes the balloon the other way – just like a rocket!

Up, Up, . . .
How far can you throw a ball? About 50 or 60 feet (15 or 18 m)? It doesn't go on forever because the Earth's gravity pulls it back down again.

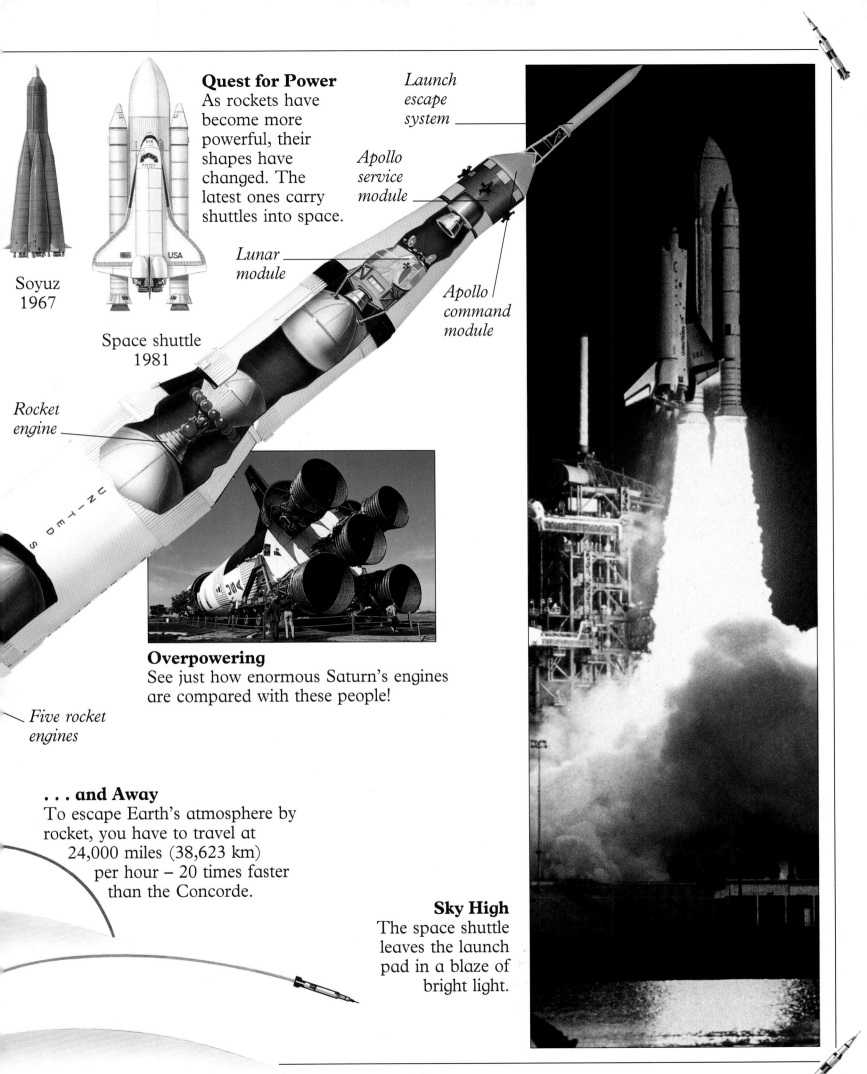

Quest for Power
As rockets have become more powerful, their shapes have changed. The latest ones carry shuttles into space.

Soyuz 1967

Space shuttle 1981

Rocket engine

Five rocket engines

Launch escape system

Apollo service module

Lunar module

Apollo command module

Overpowering
See just how enormous Saturn's engines are compared with these people!

. . . and Away
To escape Earth's atmosphere by rocket, you have to travel at 24,000 miles (38,623 km) per hour – 20 times faster than the Concorde.

Sky High
The space shuttle leaves the launch pad in a blaze of bright light.

Space Wear

THERMAL MICROMETEOROID GARMENT

TMG SHELL

In space, astronauts must wear special suits, made from many layers of material. They wear space suits when they work outside the spacecraft, because there is no air to breathe, and it can get very hot or very cold. They also wear them when they go to the Moon.

Pressure helmet

Space suits protect astronauts from radiation, keep the right pressure on their bodies, and supply them with air.

Communication connection

NASA

W. ANDERS

Penlight pocket

Oxygen supply connection

Sunglasses pocket

Carbon dioxide outlet

Space glove

Detachable pocket

The outer padded oversuit protects astronauts from dangerous dust particles.

Lunar overshoe

High and Dry
Astronauts can go to the bathroom because the space suit has a kind of diaper for women and a pouch with a tube for men.

Sensibly Suited

The astronaut wears special water-cooled underwear under the space suit.

Sun visor

"Snoopy" cap with earphones

Microphone

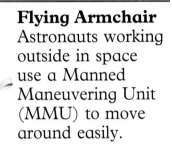

Flying Armchair
Astronauts working outside in space use a Manned Maneuvering Unit (MMU) to move around easily.

The space suit is made up of two parts: a torso (top) and a lower torso (bottom). The astronaut puts the bottom on first.

With arms up, the astronaut slides into the top half while it is still hanging up.

Hand control

The two halves are connected and locked together. Finally, the "Snoopy" cap and helmet are put on.

Nozzles, like little rockets, let out spurts of gas. These help the astronaut move and change direction.

The Trunnion Pin Attachment Device was used to grab a satellite from space to repair it, but it did not work properly. The satellite had to be caught by the remote-control arm of the shuttle.

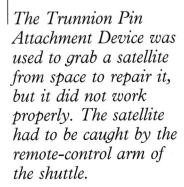

THE SPACE SHUTTLE

What space vehicle can fly into space, come back, and be used again? The answer is the space shuttle – the newest kind of spacecraft. The shuttle first flew in space in 1981. It takes off like a rocket but glides back to Earth on wings like a plane. On a space mission, which lasts about eight days, the shuttle can circle the Earth a hundred times.

After eight minutes, the fuel in the huge tank is used up, and it falls off, too.

Two minutes after liftoff, the two booster rockets fall back to Earth by parachute. They land in the sea and are picked up to be used again.

The crew crawl from the mid-deck through the tunnel to work in the Spacelab.

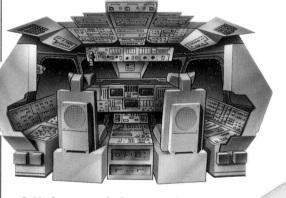

All Around Control
This is the flight deck. It has five computers and even controls on the ceiling.

Forward flight deck

Stop That Pop!
Ordinary cans would let liquids float out everywhere if opened in space, so the crew use specially made ones.

The mid-deck has a kitchen and toilet.

Vanilla instant breakfast on tray

Peanut butter

Dried pears *Dried beef*

14

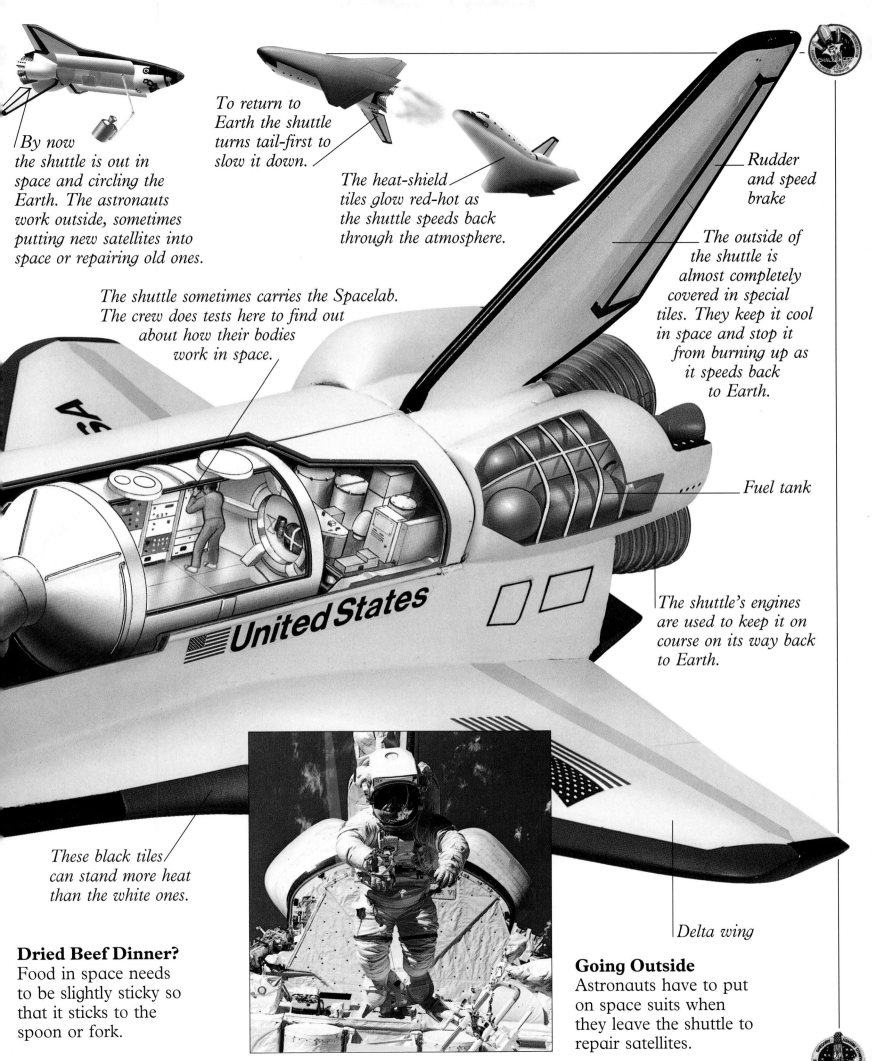

By now the shuttle is out in space and circling the Earth. The astronauts work outside, sometimes putting new satellites into space or repairing old ones.

To return to Earth the shuttle turns tail-first to slow it down.

The heat-shield tiles glow red-hot as the shuttle speeds back through the atmosphere.

Rudder and speed brake

The outside of the shuttle is almost completely covered in special tiles. They keep it cool in space and stop it from burning up as it speeds back to Earth.

The shuttle sometimes carries the Spacelab. The crew does tests here to find out about how their bodies work in space.

Fuel tank

United States

The shuttle's engines are used to keep it on course on its way back to Earth.

These black tiles can stand more heat than the white ones.

Delta wing

Dried Beef Dinner?
Food in space needs to be slightly sticky so that it sticks to the spoon or fork.

Going Outside
Astronauts have to put on space suits when they leave the shuttle to repair satellites.

15

SPACE STATIONS

Skylab

People can stay in space for a long time by living in a space station, which is a large spacecraft circling the Earth. American space travelers are called astronauts. Russian space travelers are called cosmonauts. The Russian space station is called *Mir*, the American one was *Skylab*. In *Mir*, the cosmonauts do science experiments and learn about how to live in space. Fresh water, food, and also books, letters, and videos are sent up by unmanned spacecraft.

Solar panel

Stand-up Bedroom!
Because there is no up or down in space, it is possible to sleep standing up!

The Soyuz TM *spacecraft is docking with* Mir.

As many as six spacecraft could dock here at the same time.

Soyuz TM

Station control console

Tight Squeeze
In their bulky suits, the cosmonauts in the *Soyuz TM* spacecraft do not have much room to get ready to move into space station *Mir*.

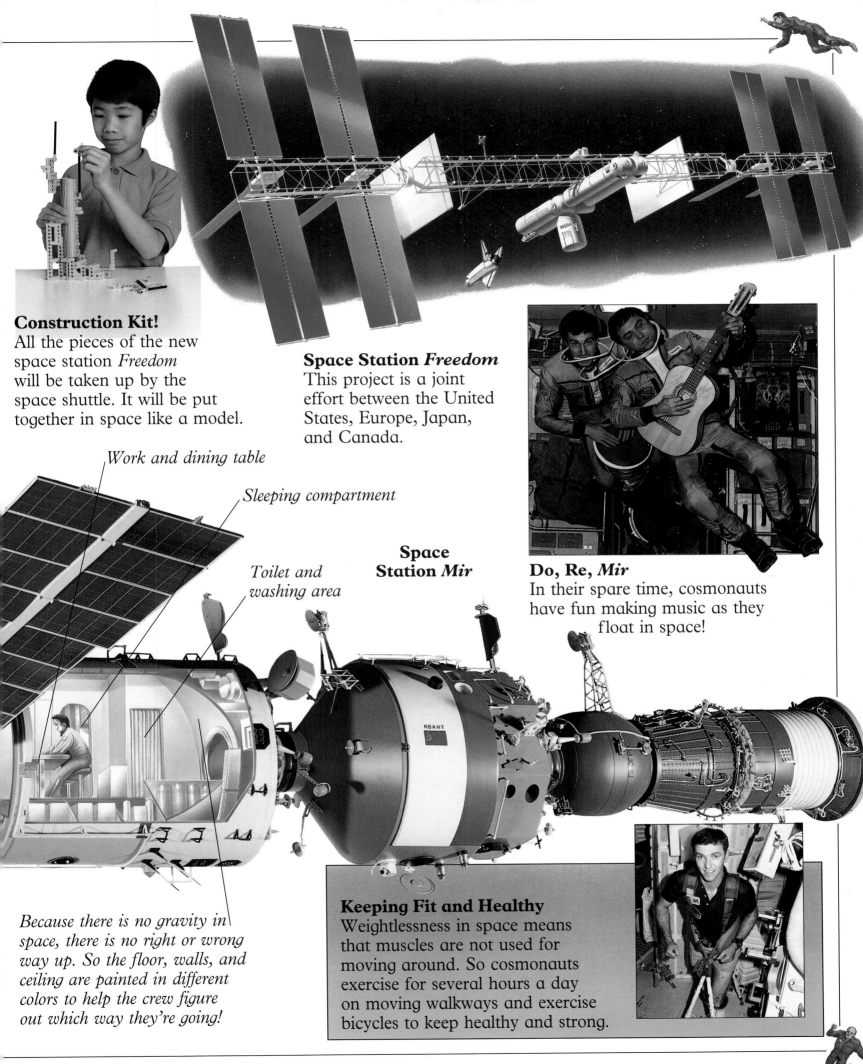

Construction Kit!
All the pieces of the new space station *Freedom* will be taken up by the space shuttle. It will be put together in space like a model.

Space Station *Freedom*
This project is a joint effort between the United States, Europe, Japan, and Canada.

Work and dining table

Sleeping compartment

Toilet and washing area

Space Station *Mir*

Do, Re, *Mir*
In their spare time, cosmonauts have fun making music as they float in space!

Because there is no gravity in space, there is no right or wrong way up. So the floor, walls, and ceiling are painted in different colors to help the crew figure out which way they're going!

Keeping Fit and Healthy
Weightlessness in space means that muscles are not used for moving around. So cosmonauts exercise for several hours a day on moving walkways and exercise bicycles to keep healthy and strong.

SATELLITES

A satellite is an object in space that orbits, or goes around, a larger object, such as a planet. The Moon is the Earth's natural satellite, but now the Earth has lots of artificial satellites as well. Satellites have to be taken into space, by rocket or by the space shuttle, and released at the right speed to stay in orbit. They come in all sorts of shapes and sizes.

Signals and messages cannot be sent in a straight line from one country to another because of the curve of the Earth, so satellites above the Earth are used to make the connections.

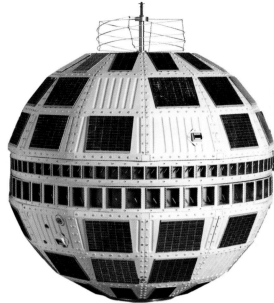

TV by *Telstar*
This satellite, called *Telstar*, sent the first live television pictures across the Atlantic in 1962.

***European Communications
Satellite 1 (ECS1)***

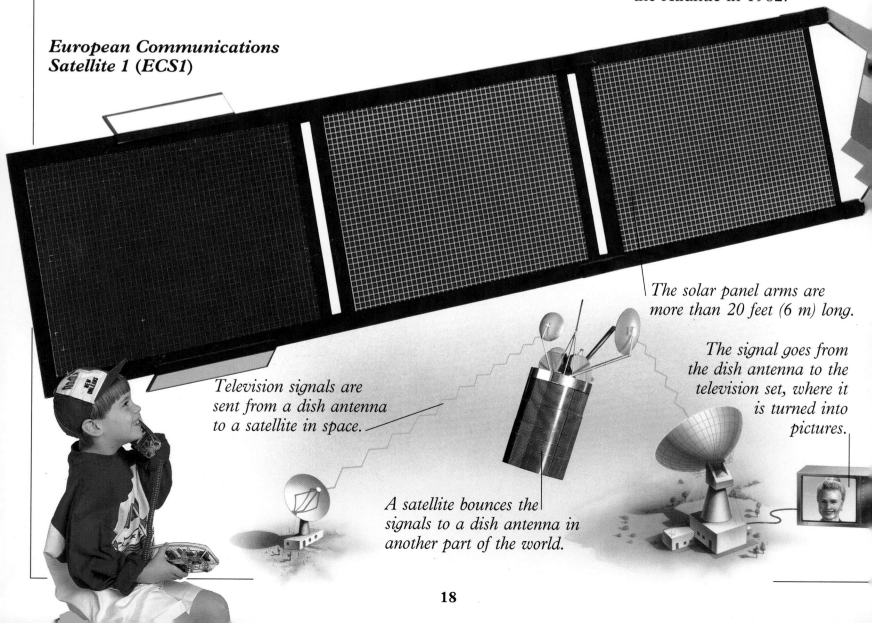

The solar panel arms are more than 20 feet (6 m) long.

The signal goes from the dish antenna to the television set, where it is turned into pictures.

Television signals are sent from a dish antenna to a satellite in space.

A satellite bounces the signals to a dish antenna in another part of the world.

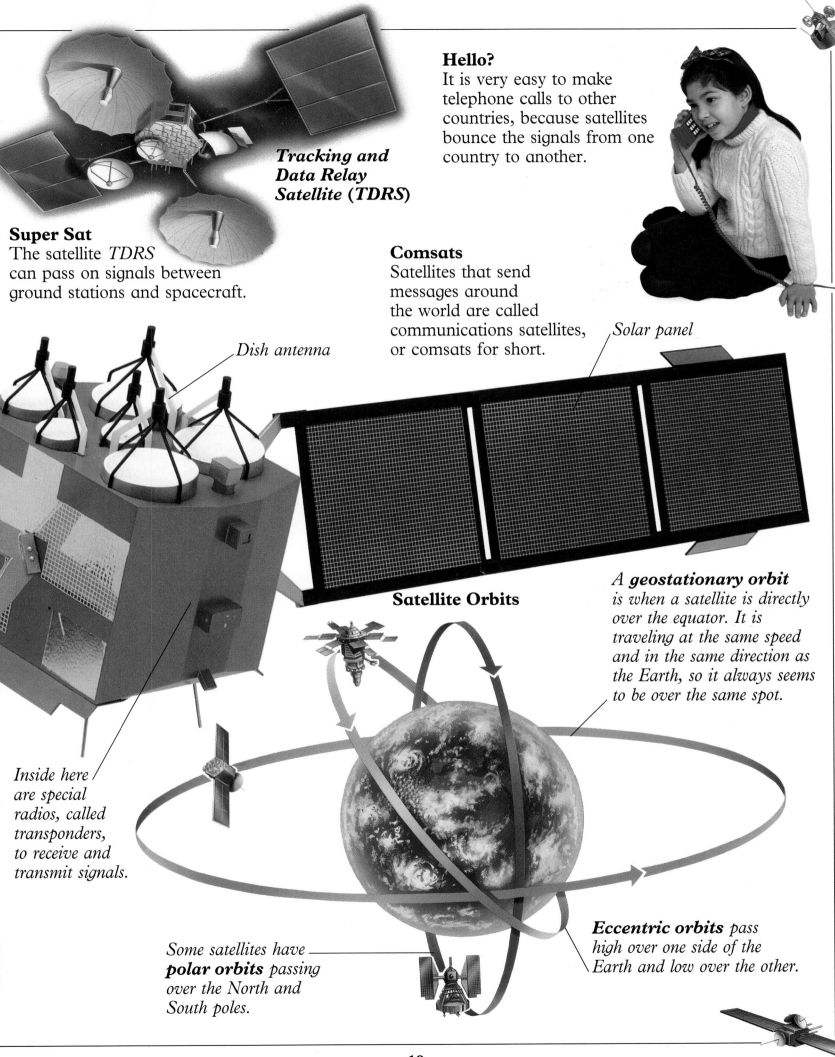

Tracking and Data Relay Satellite (TDRS)

Hello?
It is very easy to make telephone calls to other countries, because satellites bounce the signals from one country to another.

Super Sat
The satellite *TDRS* can pass on signals between ground stations and spacecraft.

Comsats
Satellites that send messages around the world are called communications satellites, or comsats for short.

Dish antenna

Solar panel

Satellite Orbits

*A **geostationary orbit** is when a satellite is directly over the equator. It is traveling at the same speed and in the same direction as the Earth, so it always seems to be over the same spot.*

Inside here are special radios, called transponders, to receive and transmit signals.

*Some satellites have **polar orbits** passing over the North and South poles.*

***Eccentric orbits** pass high over one side of the Earth and low over the other.*

SPIES IN THE SKY

Satellites travel in space, where there is no air. They often have strange looking bits and pieces that stick out. Almost all satellites are solar-powered, because the Sun always shines in space. They photograph the Earth and weather patterns, and are sometimes used to spy on people, too!

Navstar

Solar panel

Solar panel

— *Command antenna*

Mosaic Map
This Landsat mosaic shows the United States of America. The black shapes at the top are the Great Lakes.

People on the ground can tune into these navigational antennae to find out exactly where they are.

Landsat

Pictures of Our Earth
Landsat satellites, on polar orbits, take pictures of the Earth. These pictures help us do many things, such as make maps, look for oil supplies, check that crops are healthy, study groups of animals, and watch for flooding.

The cameras are under here.

Satellite dish

Lose Your Way?
Navstar's signals help ships and aircraft map their journeys. They can use its signals to find out where they are.

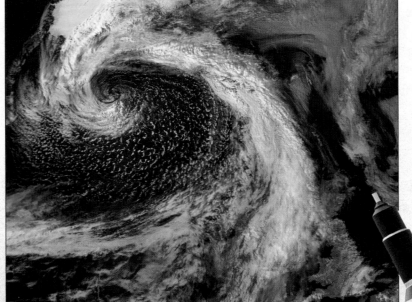

Rain Tomorrow?
Every day, weather satellites take photographs of Earth. These are sent to ground stations, where computers help people figure out what kind of weather to expect, including floods and hurricanes.

Radio antenna

Storm Clouds
The swirling clouds here show a storm brewing. The British Isles can be seen in the lower right-hand corner.

Big Bird

Geosat

A camera is inside this section.

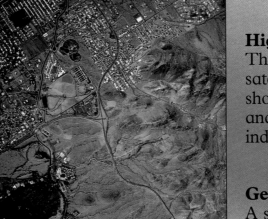

Secret film taken by a spy satellite falls to Earth in a capsule that drops off the satellite.

An airplane catches the capsule in midair and delivers it safely. If the information had been sent by radio, the enemy could have listened.

High Spy
This is a spy satellite picture showing the buildings and roads of an industrial estate.

Get the Message?
A spy satellite camera can see the print on a newspaper on Earth from as high as 36 miles (58 km). More often now, the spy capsule is not dropped, but the messages are sent using complex codes.

MOON MISSION

The second stage drops off when its five engines run out of fuel.

The command and service modules turn, join on to the lunar module, and pull it out of the third stage.

The Moon is the Earth's nearest neighbor in space, but it still takes three days to get there by rocket. It would take 200 days by car! When astronauts first went to the Moon, no one knew if it would be safe to land there. But American astronauts have been to the Moon on Apollo missions six times, and they all returned safely to Earth. The first Moon trip was in 1969 and the last in 1972.

Second stage

Astronauts traveling to the Moon crawl through a tunnel from the command module to the lunar module.

We Have Liftoff!
The first stage of the Saturn 5 rocket has five huge engines. When these run out of fuel, they fall back to Earth. Then the second stage takes over.

This air recycling unit keeps the air fresh in the cabin.

Lunar Module

CSM = Command and Service modules
LM = Lunar module
CM = Command module

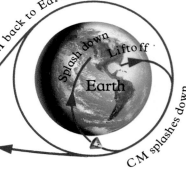

CSM docks with LM

CSM in orbit

LM back to CSM

CSM back to Earth

Splash down

Liftoff

Earth

CM splashes down

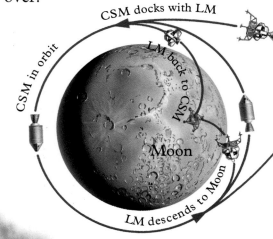

Moon

LM descends to Moon

Moon Trail
The Apollo mission to the Moon followed a path in the shape of the figure eight.

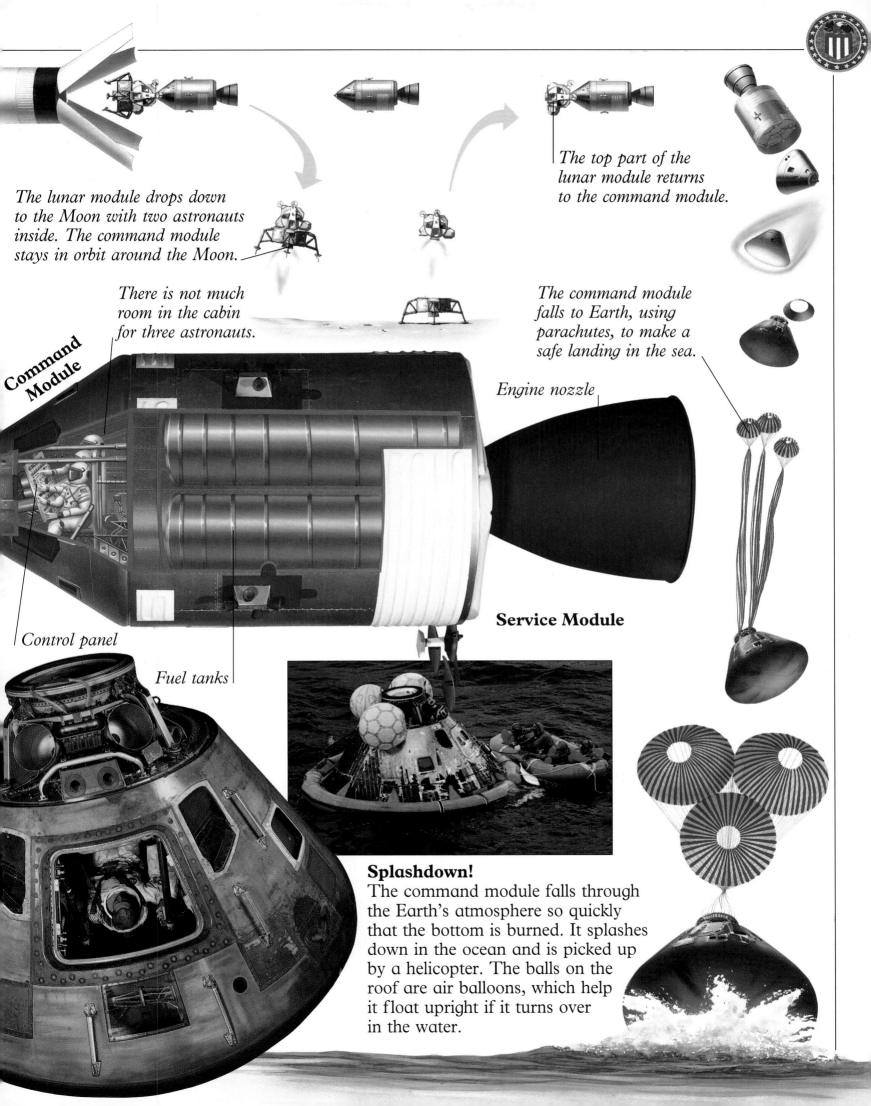

The lunar module drops down to the Moon with two astronauts inside. The command module stays in orbit around the Moon.

The top part of the lunar module returns to the command module.

There is not much room in the cabin for three astronauts.

The command module falls to Earth, using parachutes, to make a safe landing in the sea.

Command Module

Engine nozzle

Control panel

Fuel tanks

Service Module

Splashdown!

The command module falls through the Earth's atmosphere so quickly that the bottom is burned. It splashes down in the ocean and is picked up by a helicopter. The balls on the roof are air balloons, which help it float upright if it turns over in the water.

LUNAR LANDING

A lunar landing is a Moon landing. If you went to the Moon, you would find no living things at all, no air, and no water. If you stayed for a lunar "day"– about 28 Earth days – you would have two weeks of baking sun followed by two weeks of freezing nights. The first people on the Moon went down in the lunar module, known as *Eagle*.

The Apollo 11 Crew
Neil Armstrong and Edwin "Buzz" Aldrin were the first men to walk on the Moon. Michael Collins stayed in orbit in the command module.

Hanging Out the Wash?
No, just setting up a panel to collect dust! The Moon is covered in dusty soil and scattered rocks.

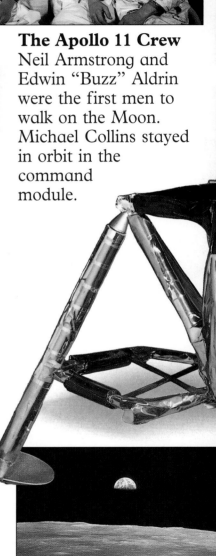

Antenna

Control panel

Television camera

Hand control

Sample collection bags

Seats

Moon Buggy
This open car was taken to the Moon for the first time in Apollo 15. Its correct name is the lunar roving vehicle.

Space for storing equipment

Wire-mesh wheel

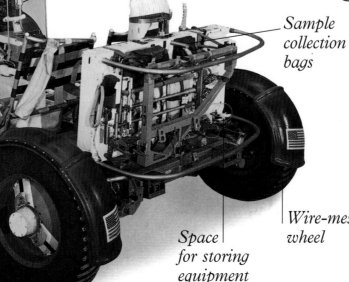

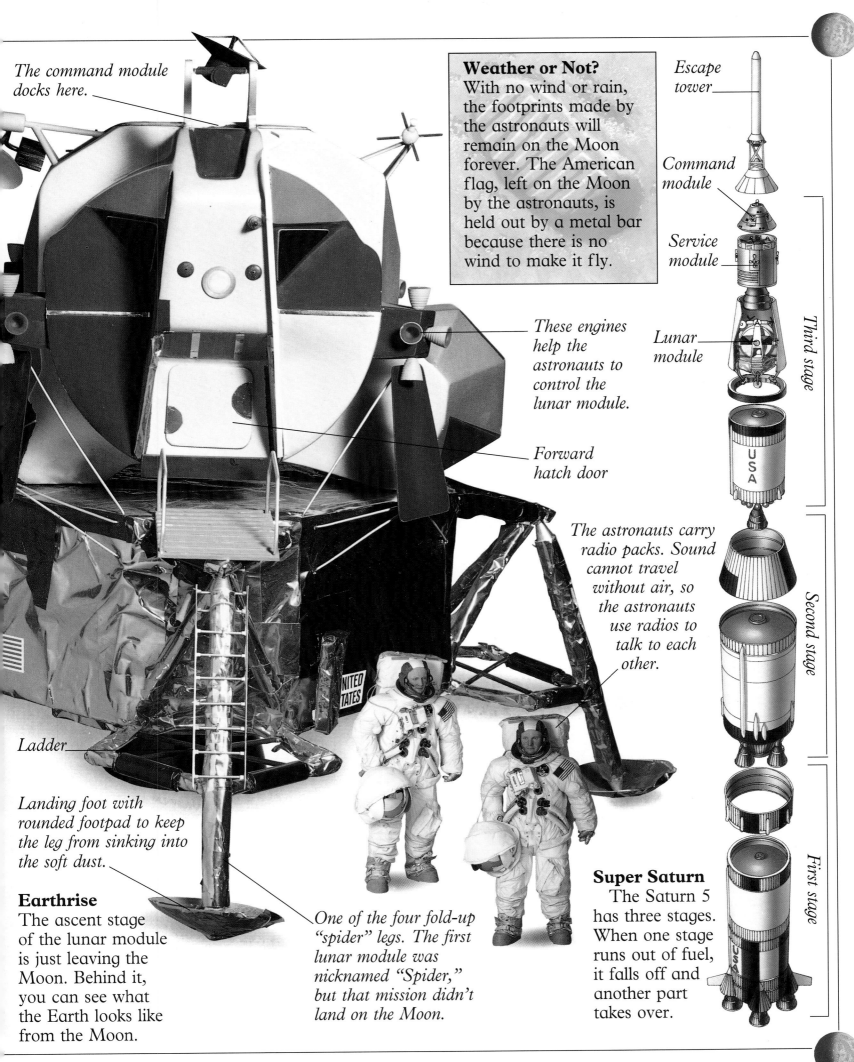

The command module docks here.

Weather or Not?
With no wind or rain, the footprints made by the astronauts will remain on the Moon forever. The American flag, left on the Moon by the astronauts, is held out by a metal bar because there is no wind to make it fly.

These engines help the astronauts to control the lunar module.

Forward hatch door

Escape tower

Command module

Service module

Lunar module

Third stage

The astronauts carry radio packs. Sound cannot travel without air, so the astronauts use radios to talk to each other.

USA

Second stage

Ladder

First stage

Landing foot with rounded footpad to keep the leg from sinking into the soft dust.

Earthrise
The ascent stage of the lunar module is just leaving the Moon. Behind it, you can see what the Earth looks like from the Moon.

One of the four fold-up "spider" legs. The first lunar module was nicknamed "Spider," but that mission didn't land on the Moon.

Super Saturn
The Saturn 5 has three stages. When one stage runs out of fuel, it falls off and another part takes over.

THE SOLAR SYSTEM

This shows the order of the planets and their positions from the Sun. It does not show their sizes.

The word "solar" means belonging to the Sun. The Sun is the center of a family of planets called the solar system. Nine planets and their moons move around the Sun in huge, oval-shaped orbits. At the same time they all spin around like tops. The four inner planets are like rocky balls, the outer ones are liquid or gas, except icy Pluto. The Earth is one of the inner planets. Without the heat and light from the Sun, there would be no life on Earth.

Prominence

Core

Radiative layer

Chromosphere

Photosphere

Corona

Sun Burn
If you could cut a slice of the Sun, you would see the core and the layers of burning gas around it.

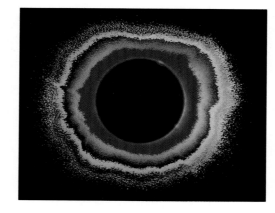

Layers of Gas
This photograph shows the layers of gas around the Sun. Added colors make the layers easier to see. This photo was taken from Earth and shows a solar eclipse, which happens when the Moon seems to cover the Sun.

SATURN

Sun Snaps
The space probe *Ulysses* will fly over the Sun and take pictures in about 1995. A space probe has no people on it.

Ulysses

NEPTUNE

URANUS

PLUTO

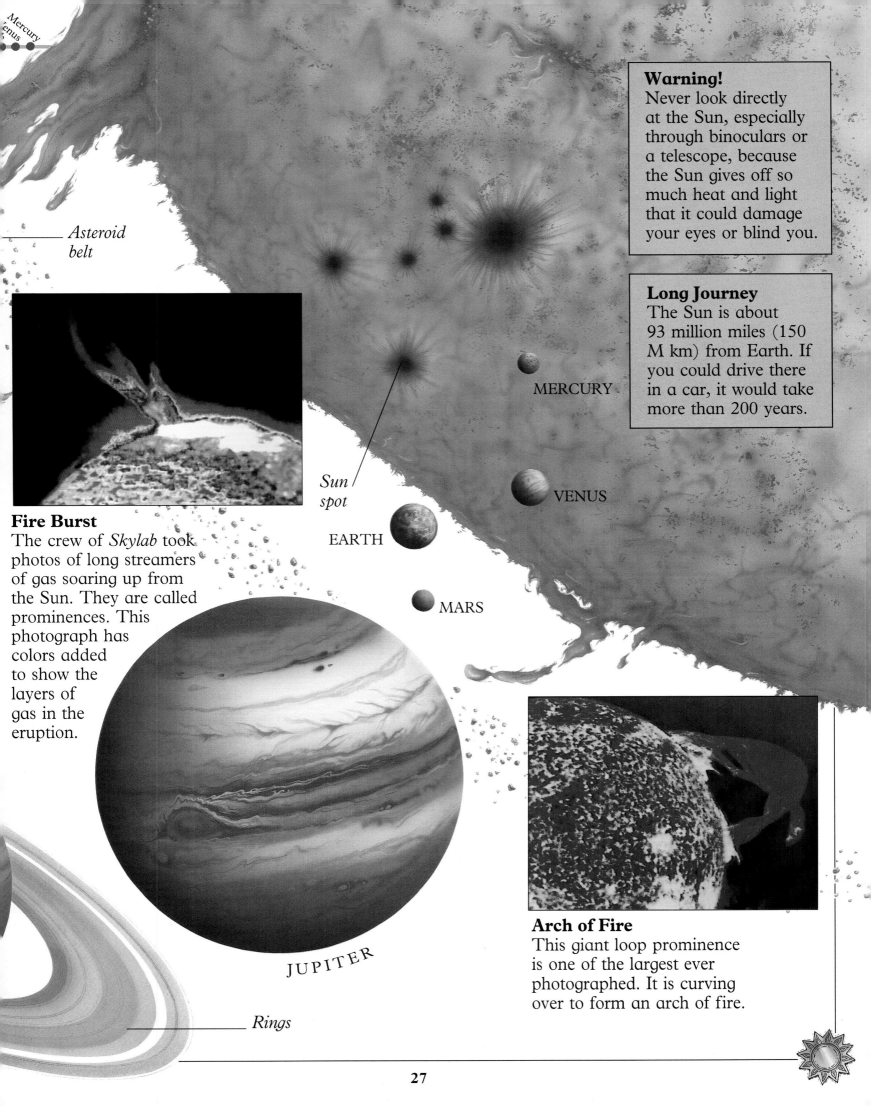

Asteroid belt

Warning!
Never look directly at the Sun, especially through binoculars or a telescope, because the Sun gives off so much heat and light that it could damage your eyes or blind you.

Long Journey
The Sun is about 93 million miles (150 M km) from Earth. If you could drive there in a car, it would take more than 200 years.

MERCURY

VENUS

Sun spot

EARTH

MARS

Fire Burst
The crew of *Skylab* took photos of long streamers of gas soaring up from the Sun. They are called prominences. This photograph has colors added to show the layers of gas in the eruption.

JUPITER

Rings

Arch of Fire
This giant loop prominence is one of the largest ever photographed. It is curving over to form an arch of fire.

MERCURY AND VENUS

Between the Earth and the Sun are two planets called Mercury and Venus. They are very hot because they are the Sun's nearest neighbors. Venus is the brightest object in the night sky, while Mercury is the second smallest planet. Photographs from space probes tell us about these planets.

MERCURY

Crust

Iron core

Magellan

Hard Center
If you could slice Mercury as you would a peach, you would find a core made of iron.

Mariner 10

Antenna

Solar panel

The solar detector made sure that the solar panels were always facing the Sun.

Television cameras sent pictures back to Earth.

Dish antenna

Star detector

Hello, Goodbye!
Mariner 10 was the first probe to visit two planets in turn. *Mariner 10* worked for 17 months before breaking down. It is now in orbit around the Sun.

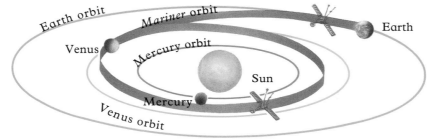

Earth orbit

Mariner orbit

Venus

Mercury orbit

Earth

Sun

Mercury

Venus orbit

The Journey of *Mariner 10*

Happy New Year!
Mercury travels fast through space and is the closest planet to the Sun. The Earth orbits the Sun every 365 days – one Earth year. Mercury's year is 88 Earth days.

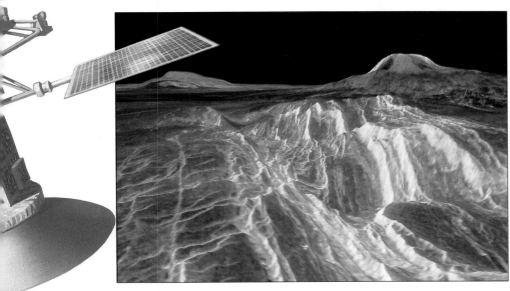

Venera 9 Venus Landing

The space probe was in a capsule on the Venera *spacecraft.*

The capsule fell through the atmosphere of Venus.

The heat-shield covers separated and fell off.

Wish You Were Here?
Magellan used radar cameras to take pictures through the thick fog around Venus. Computers made this 3-D picture of the volcanoes.

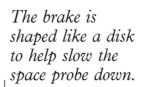

The brake is shaped like a disk to help slow the space probe down.

The probe was slowed down by a small parachute.

Venus *Venera*
Several Russian *Venera* spacecraft have been to Venus. They sent pictures back to Earth. This is the part of *Venera 9* that went down to Venus by parachute.

Instrument container

The landing ring helped make the landing soft.

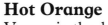

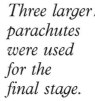

Three larger parachutes were used for the final stage.

Hot Orange
Venus is the hottest planet of all – so hot that it could melt lead! It has a bright orange sky with flashes of lightning. The Earth spins around once every 24 hours but Venus spins very slowly – once every 244 Earth days!

After a safe landing, the television cameras and instruments were switched on.

THE RED PLANET

***Viking* spacecraft**

The Viking *lander is folded into a capsule on the spacecraft.*

The red planet is Mars. It is called the red planet because its soil and rocks are red. Light winds blow the dust around, which makes the Martian sky look pink. People once thought there was life on Mars, but no life has been found so far.

It leaves the orbiter and begins its journey down to Mars.

Two *Viking* spacecraft, controlled from Earth, have visited Mars to find out what it is like. Perhaps one day people from Earth may go to live there, because it is the planet most like our own.

The television camera takes a series of pictures as it moves around.

It moves so fast that it gets very hot.

A parachute is used to slow it down. Then the heat shield drops off.

This remote-control arm is used to collect samples of Martian soil.

Tight Fit

The *Viking* lander fits into a capsule on the spacecraft. With its legs folded up, it looks a bit like a tortoise inside its shell.

The legs unfold, and the rockets are used as brakes for a soft landing.

The *Viking* lander

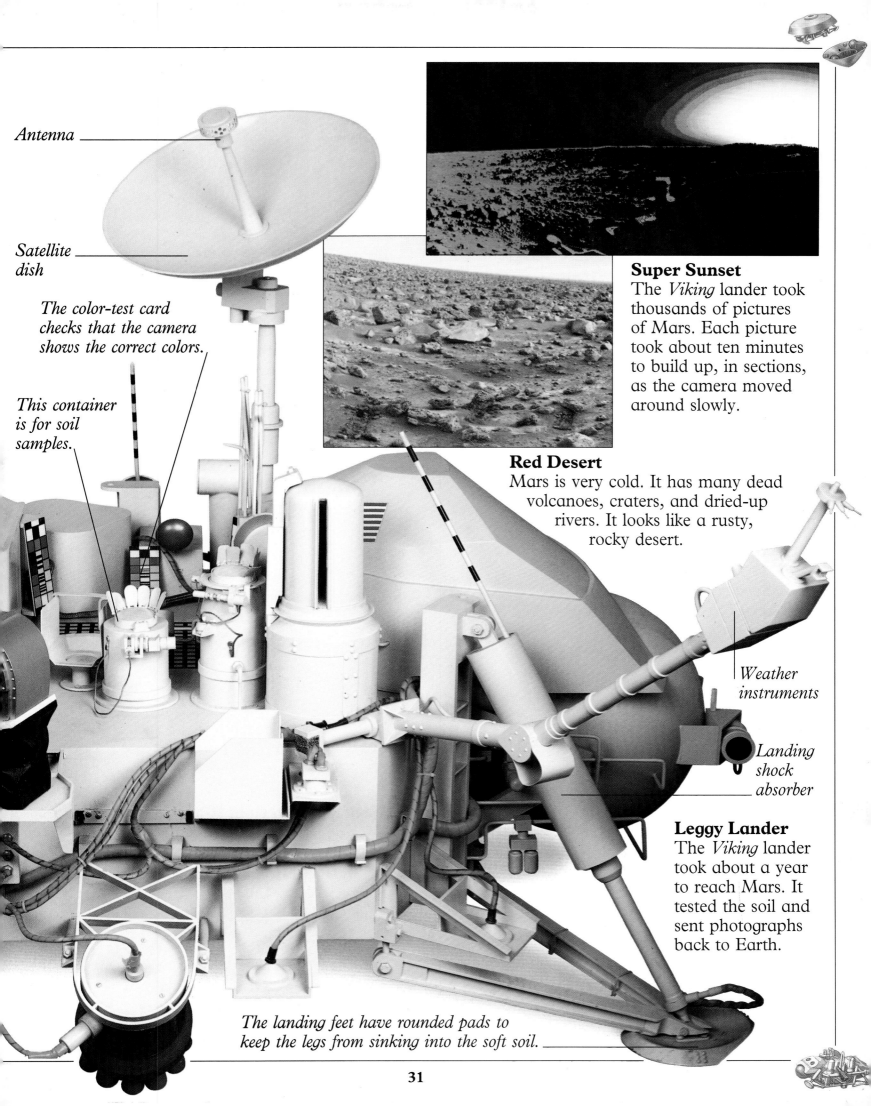

Antenna

Satellite dish

The color-test card checks that the camera shows the correct colors.

This container is for soil samples.

Super Sunset
The *Viking* lander took thousands of pictures of Mars. Each picture took about ten minutes to build up, in sections, as the camera moved around slowly.

Red Desert
Mars is very cold. It has many dead volcanoes, craters, and dried-up rivers. It looks like a rusty, rocky desert.

Weather instruments

Landing shock absorber

Leggy Lander
The *Viking* lander took about a year to reach Mars. It tested the soil and sent photographs back to Earth.

The landing feet have rounded pads to keep the legs from sinking into the soft soil.

JUPITER AND SATURN

These two giants are the largest planets in our solar system. Jupiter is made of liquid, so it is not solid enough to land on, but if you could drive a car around its equator, it would take you six months of nonstop traveling. A similar journey around the Earth's equator would take only two weeks. Saturn is a beautiful planet with shining rings around its middle. Both planets spin around very fast, pulling the clouds into stripes.

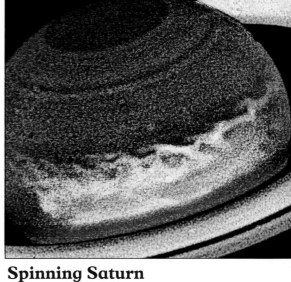

Spinning Saturn
Saturn is a giant, spinning ball of liquid held together by gravity. This photograph shows a band of clouds and the rings.

SATURN

A power supply is carried on the probe. It does not use solar power because it is working so far from the Sun.

Radio antenna

This disk has pictures of the Earth and sounds, such as a baby crying and music. If aliens find the disk, it will tell them about Earth.

Dish antenna

Seven Cold Rings
Saturn's rings are made up of glittering pieces of ice like trillions of snowballs.

Television cameras

Voyager Voyages
The *Voyager* space probes sent back pictures of Saturn and its rings.

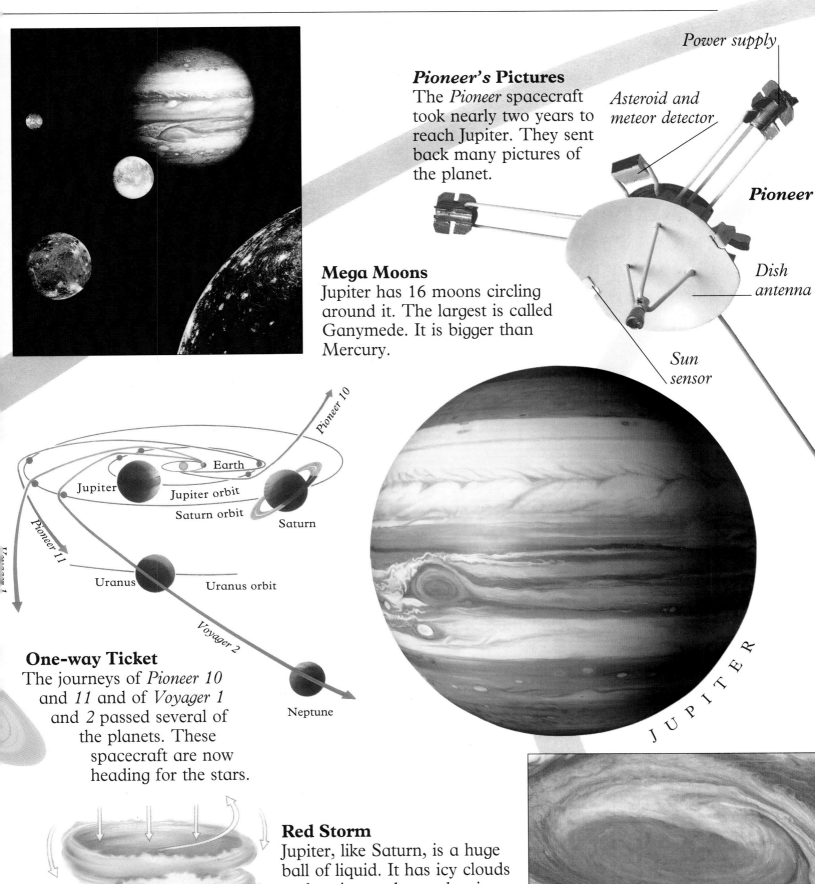

Pioneer's **Pictures**
The *Pioneer* spacecraft took nearly two years to reach Jupiter. They sent back many pictures of the planet.

Power supply

Asteroid and meteor detector

Pioneer

Dish antenna

Mega Moons
Jupiter has 16 moons circling around it. The largest is called Ganymede. It is bigger than Mercury.

Sun sensor

Pioneer 10

Earth

Jupiter

Jupiter orbit

Saturn orbit

Saturn

Pioneer 11

Uranus

Uranus orbit

Voyager 2

J U P I T E R

One-way Ticket
The journeys of *Pioneer 10* and *11* and of *Voyager 1* and *2* passed several of the planets. These spacecraft are now heading for the stars.

Neptune

Red Storm
Jupiter, like Saturn, is a huge ball of liquid. It has icy clouds and a giant red spot that is the center of a huge storm.

Swirling winds blow Jupiter's clouds into a hurricanelike storm.

THE OUTER PLANETS

Sideways Spinner
Uranus looks as if it is spinning on its side. It is covered in dense fog.

Uranus, Neptune, and Pluto are the farthest planets from the Sun, so they are called the outer planets. They are very cold. Uranus was the first planet to be discovered using a telescope, because you cannot see it from Earth just with your eyes. Pluto is the farthest away, and no spacecraft has visited it yet. If a jet could fly there, it would take a thousand years!

1986UIR
Epsilon
Gamma
Delta
Eta
Alpha
Beta
4 5 6

Outer Solar System

The rings of Uranus are made of rocks. The widest ring, called Epsilon, is 58 miles (93 km) across.

Orbit of Pluto

Ellipses
The planets move around the Sun in squashed circles called ellipses. This girl is drawing ellipses.

Order of Orbits
"Planet" means "wanderer." The planets travel around the Sun in paths, called orbits. The ones nearer the Sun have shorter lengths of orbit than the ones farther away. Some astronomers believe there is a tenth planet, not yet discovered. They call it Planet X and it may be bigger than Pluto.

This diagram shows the orbits of the planets and their places in the solar system. It does not show the correct sizes of the planets.

NEPTUNE

PLUTO

Blue Neptune
The *Voyager* spacecraft took photographs of Neptune's clouds and its storm, which is called the Great Dark Spot.

Inner Solar System

Brrrrr!
Pluto is tiny – smaller than our Moon – and at night is ten times colder than a freezer!

Light from the Sun takes three minutes to reach Mercury and eight minutes to reach the Earth, but five and a half hours to reach Pluto!

The Sun
Orbit of Mercury
Orbit of Venus
Orbit of Earth
Orbit of Mars

Pluto is not always the farthest planet from the Sun. Its strange orbit means that sometimes it is nearer than Neptune.

Orbit of Jupiter
Orbit of Saturn
Orbit of Uranus
Orbit of Neptune

ON THE MOVE

Between Mars and Jupiter, there is a belt of rocks in orbit called the asteroid belt. The chunks of rock are called asteroids. Sometimes these rocks crash into each other and bits fall down toward Earth.

Also in orbit around the Sun are lumps of rock and ice, called comets. When comets get near the Sun, they shine like "hairy stars." This is what people used to call them long ago. The most famous comet is Halley's comet, named after the man who first studied it.

Path of Halley's comet

Wind from the Sun blows the dust and gas around Halley's comet into an enormous tail. Comets' tails always point away from the Sun and can be millions of miles long.

Solar system

Halley's comet

ISTI MIRANT STELLA

HAROLD

Earth Comets
Because they are made of rock and ice, comets are often called "dirty snowballs." Make your own comet the next time it snows!

Regular Visitor
We see Halley's comet from Earth once every 76 years because it takes that long to orbit the Sun. It was shown in a picture called the Bayeux Tapestry more than 900 years ago!

The comet's center is made of ice. As it gets near the Sun, the ice melts.

A Belt You Cannot Wear
The asteroids in the asteroid belt are really mini-planets. There are thousands of them. The largest is about 600 miles (1,000 km) across.

Comet Head
This is a photograph of the head of Halley's comet. Computer colors show the bright center and the layers around it.

Meteor Shower

If a lump of rock or metal burns up before it reaches the ground, it is called a meteor or shooting star. This photograph shows lots of them falling together in a meteor shower.

Crash! Bang!

A large meteor that does not burn up as it plunges through the Earth's atmosphere is called a meteorite.

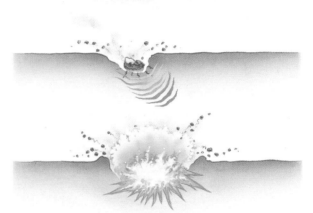

It travels so fast, it shatters into pieces as it hits the ground.

It causes shock waves as it lands.

The explosion leaves a big hole, called a crater.

This huge meteorite crater is in Arizona.

Dish antenna

Solar cells for power

Giotto

Camera

Dust shield

Comet Quest

In 1986, *Giotto*, the European spacecraft, passed very close to Halley's comet and took pictures of it. *Giotto* is now searching for other comets.

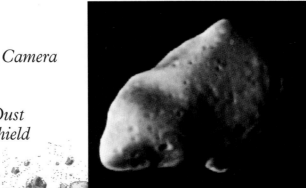

Gaspra the Asteroid

No one had seen a picture of Gaspra until the spacecraft *Galileo* took this one in 1991, as it flew past the asteroid belt.

SKY WATCHING

If you look up at the sky on a clear night, you can see hundreds of stars and, sometimes, the Moon. But if you use binoculars or a telescope, you can see even more – for example, the planets and the craters on the Moon.

When astronomers study the universe they use huge radio telescopes, some with dishes, to help them see far, far away, and to gather information from space. The Hubble Space Telescope is the largest telescope to be put into space. It can provide clear pictures of stars and galaxies because it orbits 380 miles (612 km) above the Earth's murky atmosphere.

Clearly Venus
This photograph of Venus was taken by the *Pioneer* Venus orbiter. It used radar to get a clear picture through the thick clouds around Venus. The signals were sent back to Earth to a radio telescope, where this picture was produced.

Solar panel

Radio telescope

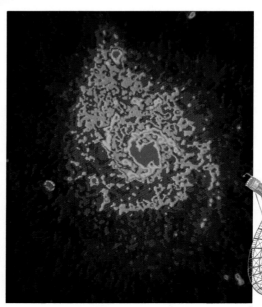

Whirligig
This radio map of the Whirlpool galaxy was taken by a radio telescope. The added colors show the spiral arms of the galaxy.

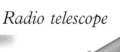

Head in the Stars
Through a telescope you can see the Horsehead nebula. This picture has false colors added, but it looks nearly as bright without them.

Star Belt
This photograph, taken through a small telescope, shows part of the constellation of Orion (also called "The Hunter").

Flap door

Star Cluster
This photograph was taken by the Hubble Space Telescope. It shows a star cluster.

Small mirror

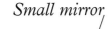

Main mirror

Starlight
This natural-color photograph was taken from an observatory. It shows the Orion nebula, which is a cloud of dust and gas lit from inside by newly born stars.

Double Hubble
The Hubble has two mirrors. The largest is 7.9 feet (2.4 m) wide and 12 inches (30 cm) thick. The mirrors are not working as well as they should, but astronauts from the shuttle will be correcting them soon.

Antenna

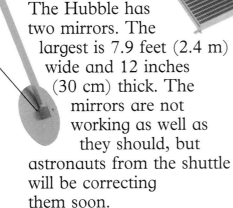

Lookout
An observatory is a place where astronomers work. These are usually away from big cities, in places where there are no streetlights and the air is clear.

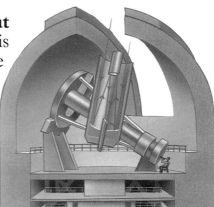

STARS AND GALAXIES

Stars look like tiny points of light from the Earth but really they are huge, hot balls of burning gas deep in space. They are forming, changing, and dying all the time. There are big stars called giants, even bigger ones called supergiants, and small ones called dwarfs. Our Sun is just one of about a hundred thousand million stars that all belong to a galaxy called the Milky Way. A galaxy is a group of millions of stars, held together by a strong force called gravity.

Starry, Starry Night
On a clear night, do not forget to look up at the sky! You will see hundreds of twinkling stars, like tiny sparkling diamonds, far above you.

The gas and dust pack tightly together, getting smaller and very hot.

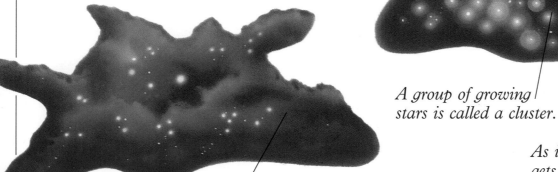

A group of growing stars is called a cluster.

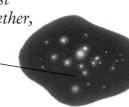

A new star is very bright. It shines steadily for many years.

As it cools, the star gets bigger and forms a red giant.

A star is born inside a huge cloud of dust and gas, called a nebula. The word "nebula" means "mist."

Near the end of its life, the core of a red giant may cave in and give off layers of gas.

Sky Lights
There are many new stars in the gas and dust of this pink nebula. A new young star on its own shines blue.

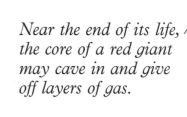

Sometimes a giant star explodes and is blown to pieces. This is called a supernova.

A supernova explosion sometimes results in a pulsar. It spins very fast and sends out sweeping beams of radio waves.

A black hole is not really a hole, but a very tightly packed object. It is solid and does not reflect any light, so it looks like a hole! Its gravity pulls things toward it like water down a drain.

Some dying stars grow into huge red supergiants.

Our solar system

Galaxies

A spiral galaxy

An elliptical galaxy

Massive stars shine very brightly but do not live as long as smaller stars.

If seen through a telescope, the star now looks like a planet, so it is called a planetary nebula.

Star Spinner

Our galaxy, called the Milky Way, is a spiral galaxy. Our solar system is about two-thirds of the way out from the center, in one of the spiral arms. There are lots of galaxies in the universe, and they have different shapes. Try painting some!

Some stars gradually get smaller and whiter until they become white dwarf stars.

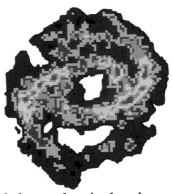

A barred spiral galaxy

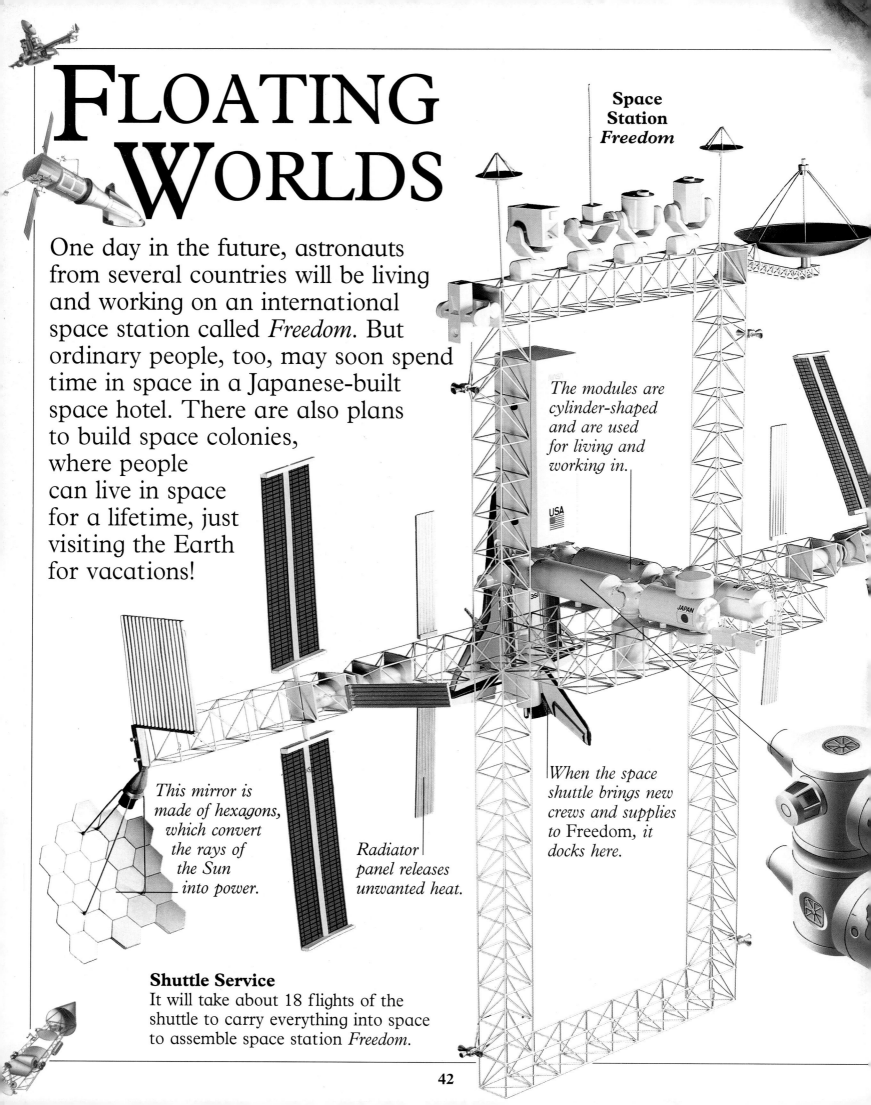

FLOATING WORLDS

Space Station *Freedom*

One day in the future, astronauts from several countries will be living and working on an international space station called *Freedom*. But ordinary people, too, may soon spend time in space in a Japanese-built space hotel. There are also plans to build space colonies, where people can live in space for a lifetime, just visiting the Earth for vacations!

The modules are cylinder-shaped and are used for living and working in.

This mirror is made of hexagons, which convert the rays of the Sun into power.

Radiator panel releases unwanted heat.

When the space shuttle brings new crews and supplies to Freedom, *it docks here.*

Shuttle Service
It will take about 18 flights of the shuttle to carry everything into space to assemble space station *Freedom*.

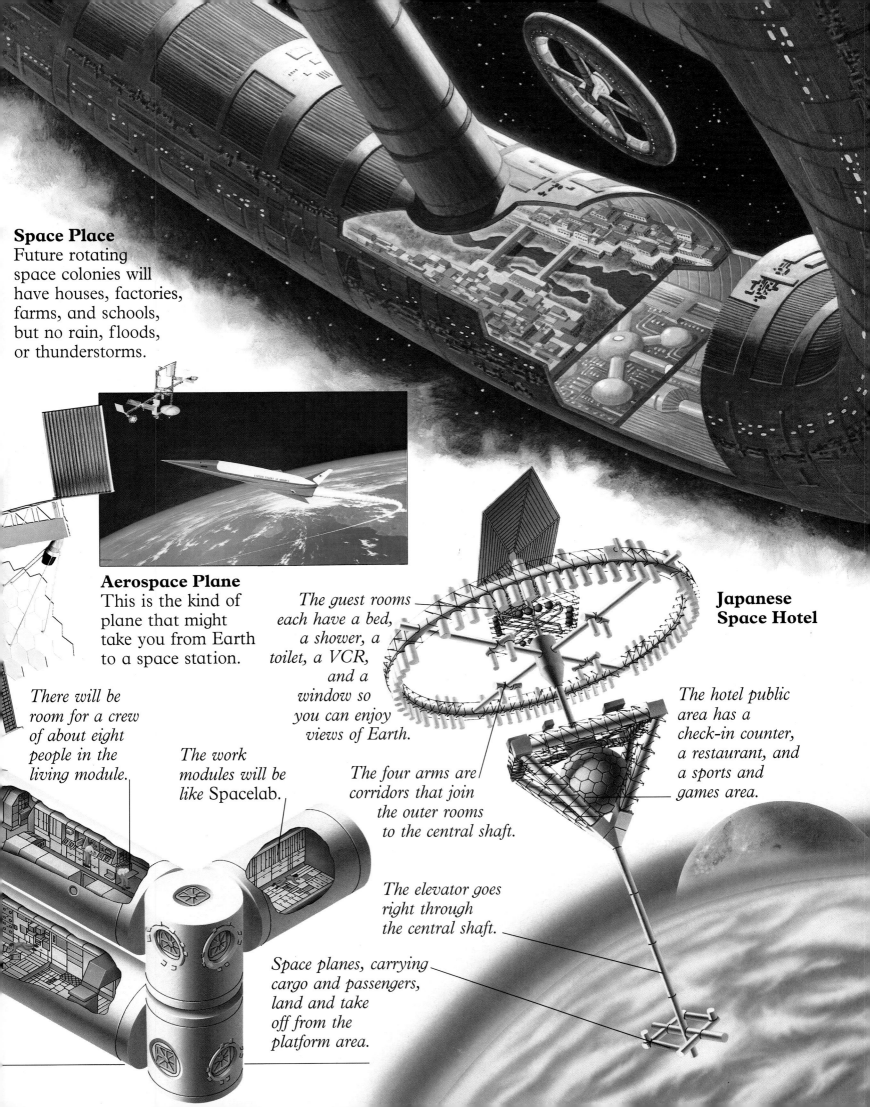

Space Place
Future rotating space colonies will have houses, factories, farms, and schools, but no rain, floods, or thunderstorms.

Aerospace Plane
This is the kind of plane that might take you from Earth to a space station.

There will be room for a crew of about eight people in the living module.

The work modules will be like Spacelab.

The guest rooms each have a bed, a shower, a toilet, a VCR, and a window so you can enjoy views of Earth.

The four arms are corridors that join the outer rooms to the central shaft.

Japanese Space Hotel

The hotel public area has a check-in counter, a restaurant, and a sports and games area.

The elevator goes right through the central shaft.

Space planes, carrying cargo and passengers, land and take off from the platform area.

LIVING IN SPACE

A home on another planet may be a dream today, but it is quite possible that one day people will be living on the Moon or on Mars. The first settlements will be quite simple, and will be built under the surface. People will have to stay inside the protective buildings or wear a space suit because of the lack of oxygen and the danger from radiation.

The first people to go will probably be scientists and astronomers. Scientists think there are useful metals to be found there, and astronomers will build huge telescopes so they can study the universe.

A Base on Mars

Solar panels will be used to convert the Sun's energy into power.

Nothing can grow on Mars unless it is in a special greenhouse. Water and air will need to be controlled.

Moon and Mars Mines
Aluminum, iron, and other useful metals will be mined. The materials that are mined will be used where they are or taken back to Earth.

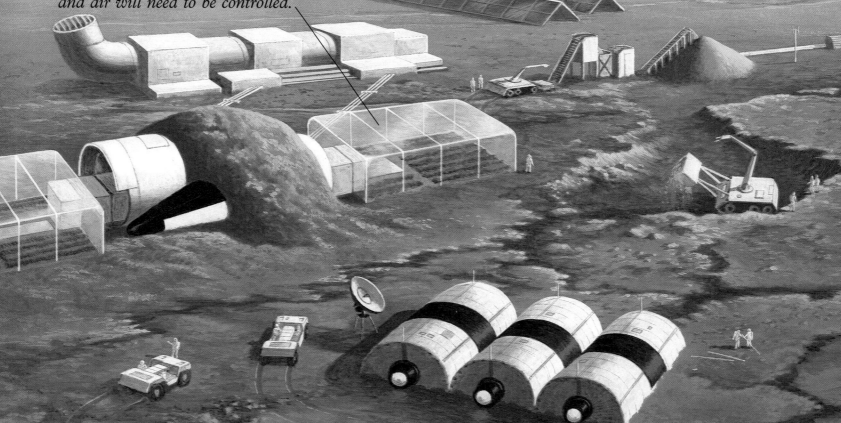

The inflatable dome is a temporary workshop used for repairing a Moon buggy.

Small moon-landers can be used as computer rooms.

Solar panels convert the energy from the Sun into electricity. Any extra electricity that is collected will be stored and can be used at night.

The Mars landing craft will not be wasted. They will be turned on their sides and used as Martian houses and workplaces.

Moon City
A team goes out to explore the surface of the Moon. A large city has already been built.

Dust Gliding
Unlike the Moon, Mars has dust storms, which give the sky a red glow. Perhaps gliders will be used for traveling in the Martian winds.

THE FUTURE

No one knows if there is life in other galaxies, or even in other parts of our own galaxy. Many people claim to have seen alien spacecraft, which are sometimes known as UFOs (Unidentified Flying Objects), or flying saucers. Some people say they have even met the aliens!

To find out if there really are any aliens, astronomers listen for radio messages from space, and even send out their own messages, hoping that one day they will get an answer!

Starship
Huge spacecraft may travel long distances from Earth through the galaxies, searching for other life.

Robot Traveler
In the future, robot vehicles will be specially made to land and move around easily on the rocks and craters of other planets.

Call the Garage
If your spacecraft broke down on another planet, it would be much too dangerous to leave the safety of your craft. Robots would be used to repair it.

Ramscoop
This is a ramscoop starship. It has been damaged on its journey and is returning to its planet for repairs. There is life on this planet – but is it human life?

The Visitor
Has this strange creature just arrived on Earth? Is the spacecraft in the sky full of friendly aliens?

Space Strangers
A planet with water might have some form of fish or insect life.

Creatures on a planet with low gravity might keep on growing and have long, skinny arms and fingers.

On a planet with a stronger pull of gravity, the creatures might be squashed toward the ground.

Floating Homes
Life in this space city would be like life on Earth, except that this city can be moved to another planet.

GLOSSARY

Antenna An aerial for transmitting or receiving radio signals.

Asteroid A small rock or metal object that orbits the Sun like a tiny planet. Most are in a belt between Mars and Jupiter.

Astronaut A person who travels beyond the Earth and into space.

Astronomer A person who studies the stars, planets, and other objects in space.

Atmosphere A blanket of gases that surrounds a planet or moon.

Barred spiral galaxy A group of stars collecting together to make a spiral shape with a bar across the center.

Command module The cone-shaped capsule in which the astronauts traveled during the Moon missions.

Comet An object in the solar system made of ice and dust, which shines as it gets near the Sun.

Constellation The pattern made by a group of stars in the night sky.

Cosmonaut The Russian word for astronaut.

Eccentric orbit When a satellite passes low over one side of the Earth and high over the other.

Eclipse When the shadow of one planet or moon falls on another.

Elliptical galaxy A huge group of stars collecting together to make an egg-like shape.

Equator The imaginary line around a planet halfway between the North and South poles.

Galaxy A huge "island" of stars in space.

Geostationary orbit When a satellite is directly over the Equator, traveling in the same direction and at the same speed as the Earth.

Gravity The force of a planet that tries to pull everything toward its center.

Hemisphere Half of a sphere. The Earth is divided into the northern and southern hemispheres by the equator.

Light-year The distance traveled by a beam of light in one Earth year.

Lunar Of or relating to the Moon.

Lunar module The craft used by the Apollo astronauts to land on the Moon.

Manned Maneuvering Unit A small jet-pack worn by astronauts so that they can move around easily in space.

Meteor A chunk of rock or metal that burns up as it falls through the Earth's atmosphere. Sometimes called a shooting star.

Meteorite A meteor that does not burn up and that reaches a planet's surface.

Moon The natural satellite of a planet.

Observatory A place where astronomers go to study the night sky.

Orbit The path taken by one object around another, such as the Earth around the Sun.

Planet A large round object that orbits a star. The Earth is one of the nine planets around the Sun. Planets do not shine, but they reflect the Sun's light.

Polar orbit When a satellite passes over the North and South poles.

Pulsar A rapidly spinning star that gives off pulses of radio waves.

Quasar A very bright, distant object, which may be the center of a faraway galaxy.

Rocket A very powerful engine to launch people and objects into space.

Satellite A small object that circles around a larger one – either natural or artificial.

Service module The part of the Saturn 5 rocket that carried the engine, fuel, and supplies for the Apollo Moon missions.

Solar Of or relating to the Sun.

Solar panel The outside part of a satellite that is covered in cells that change sunlight into electricity.

Solar system The family of the Sun including the planets, moons, asteroids, meteors, and comets.

Space probe A craft with no people in it that travels into space to explore the planets.

Space shuttle A space plane that carries astronauts to work in space and is reusable.

Space station A large structure in space where astronauts live and work for long periods of time.

Spacelab A workshop carried into space by the space shuttle.

Space suit A special suit worn by astronauts when they go spacewalking. It protects them from the dangers in space and supplies them with oxygen to breathe.

Spiral galaxy A huge group of stars together making a spiral shape.

Star A huge ball of burning gas giving out heat and light. Our Sun is a medium-sized star.

Supernova A huge star that explodes and blows itself apart.

Telescope A tube, with mirrors and lenses, to look through, which makes far-away objects look closer and bigger.

UFO Unidentified Flying Object A mysterious object in the sky that no one can explain.

Universe Everything that exists in space.

Weightlessness In space there is no gravity, so things seem to have no weight and float around.

INDEX

Acknowledgments

Photography: Tina Chambers; Geoff Dann; Steve Gorton; James Stevenson. **Illustrations:** David Bergen; Bob Corley; Tony Gibbons; Mick Gillah; Terry Hadler; Keith Hume; Chris Lyon; Sebastian Quigley; Roger Stewart; Grose Thurston; Graham Turner. **Models:** Atlas Models; Peter Griffiths. **Thanks to:** ESA; London Planetarium; The Science Museum, London; Truly Scrumptious Child Model Agency.

Picture credits

Genesis Space Photo Library: 16; **Michael Holford:** 36c; **Image Bank:** Dave Archer endpapers; **NASA:** front cover c, clb & crb, 3l, 5tr(3), 11l, 13t & b, 15, 17b, 20/21, 23, 24t, c, bl & br, 27b, 28b, 29t, 31t & b, 32t, 33t & b, 35b, 37b, 38c, 39cr, 43cl; **Science Photo Library:** 5tr(2), 26t, 39tr, Julian Baum 45c & b, Dr. Martin N. England 5b(2), 41br, Dr. Fred Espenak front cover cra, 4br, 26b, 41cr, David A. Hardy 41c, 46b, 47t & b, Kapteyn Laboratorium 38b, Dr.John Lorre 5b(1), 41ucr, NASA 3br, 27t, 35t, 36b, 44c, NASA/Dr. Gene Feldman/GSFC 20t, NOAO 40b, Mark Paternostro 47r, Max Planck Institute 38t, Ronald Royer 40/41, John Sanford 5tr(1), 8, 9, 37 t & r, 39tl, Dr. Rudolph Schild/Smithsonian Astrophysical Observatory 40t, US Naval Observatory 39br; **Tass:** 17t; **Telegraph Colour Library:** Space Frontiers 11r, 21b, Space Frontiers/N.R.S.C. 21t.

t – **top**   l – **left**   a – **above**   cb – **center below**
b – **bottom**   r – **right**   u – **under**   c – **center**   clb – **center left below**
crb – **center right below**   cra – **center right above**